THOMAS AND
THE PHYSICS
OF 1958:
A CONFRONTATION

The Aquinas Lecture, 1958

THOMAS AND THE PHYSICS OF 1958: A CONFRONTATION

Under the Auspices of the Aristotelian Society
of Marquette University

By

HENRY MARGENAU, Ph.D.

MARQUETTE UNIVERSITY PRESS
MILWAUKEE
1958

Library of Congress Catalog Card Number: 58-9679

PRINTED
IN
U. S. A.

Prefatory

The Aristotelian Society of Marquette University each year invites a scholar to deliver a lecture in honor of St. Thomas Aquinas. Customarily delivered on a Sunday close to March 7, the feast day of the society's patron saint, the lectures are called the Aquinas lectures.

In 1958 the Aquinas lecture "Thomas and the Physics of 1958: A Confrontation" was delivered on March 2, in the Peter A. Brooks Memorial Union of Marquette University, by Henry Margenau, Eugene Higgins Professor of Physics and Natural Philosophy, Yale University.

Born in Bielefeld, Germany, in 1901, Professor Margenau studied at the Teacher's College, Herford, Germany. He received his B.A. degree from Midland College, Nebraska, in 1924, his M.S. degree from the University of Nebraska, in 1926, and his Ph.D. from Yale, in 1929.

He joined the Yale faculty as instructor in physics in 1928, studied at Munich and Berlin 1929-30 on a Sterling Research Fellowship, was promoted to assistant professor at Yale when he returned, became full professor in 1946. He was appointed the first Eugene Higgins Professor of Physics and Natural Philosophy in 1950.

During World War II, Professor Margenau worked in micro-wave theory, particularly duplexing systems, the devices that make it possible to use a common antenna in radar for both sending and receiving operations. He has made important contributions to physics in his work on spectroscopy and nuclear physics, and the discharge theory.

Professor Margenau is President of the Philosophy of Science Association; a member of the governing board of the Association of Philosophy of Science and of Sigma Xi, and a trustee of Connecticut College for Women. He has been an associate editor of the *Journal of the Philosophy of Science* (1933), of the *American Journal of Science* (1943) and the *Review of*

Modern Physics (1954). He is a fellow of the American Physics Society and of the American Academy of Arts and Sciences.

He has been visiting lecturer at the University of Washington, University of California, New York University; Philips Lecturer at Haverford College; Visiting Professor at the University of Heidelberg, Germany; Hill Foundation Professor at Carleton College, Joseph Henry Lecturer at Washington, D.C. and Remsen Bird lecturer at Occidental College. In 1957 he was appointed national visiting scholar by Phi Beta Kappa.

He was honored by Michigan State University in May, 1955 when it presented him with its Centennial Award for his work in physics and philosophy. He received an honorary doctor of humane letters degree from Carleton College in 1954 and an honorary doctor of science degree from the University of Nebraska in 1957.

Professor Margenau has served as consultant to the Atomic Energy Commission, the Brookhaven National Laboratory, the

National Bureau of Standards, the Argonne National Laboratory, the Rand Corporation of California and a number of industrial firms. He has been a staff member of the Institute for Advanced Study at Princeton and the Radiation Laboratory of the Massachusetts Institute of Technology.

He is the author of *Foundation of Physics* (with R. R. Lindsay) (New York, Dover Publications, 1957); *Mathematics of Physics and Chemistry* (with G. M. Murphy) (Princeton, Van Nostrand, 1956) 2nd edition; *The Nature of Physical Reality* (New York, McGraw-Hill, 1950); *The Nature of Concepts;* and almost 150 scientific and philosophical articles in learned journals.

To these the Aristotelian Society takes pleasure in adding *Thomas and the Physics of 1958: A Confrontation.*

Thomas and the Physics
of 1958: A Confrontation

I

L AST July after attending a scientific
meeting, I found myself stranded in
Rome between planes for one glorious Sunday morning. The sun was bright and the
eternal city beckoned the weary traveler
with all its charms. Thus I decided to take
a bus to Vatican City and to refresh the
memory of previous visits to St. Peter's in
the scant two hours at my disposal. On
entering the Cathedral and looking about,
surrounded by the splendid architecture
of the basilica and the treasures of centuries, I felt oppressed and overwhelmed,
for I had neither the time nor the competence to appreciate St. Peter's in its total-
ity. My frustration was aggravated by the

noise of tourists, by the distracting activities of busybodies scurrying to and fro among the tombs and altars. Under these circumstances I made a decision which restored my equanimity: I decided to ignore the total effect, forcing myself to limit attention to a few details. I recalled, for instance, the great contributions of Michelangelo and was able to enjoy and value the architectural plan of the cross, the famous dome, the beautifully sculptured Pietà in the first small chapel near the entrance. But I saw almost nothing else.

The present occasion repeats many of the experiences of that visit to the Vatican Hill. Upon accepting the challenge to give the Aquinas lecture, I reentered the cathedral of Thomas' writings, to which I was a relative stranger. In the face of the magnificent details of his philosophic system I found myself bewildered. While undertaking a review, I was often annoyed by the effusions of pedantic commentators who made it a pleasure to return to

Thomas' own lucid texts. And so I was forced, as in St. Peter's, by lack of time and limited competence to focus attention upon but a few details of workmanship in the saint's philosophy. Selecting what appear to be relevant portions of his writings for special study, I propose to confront his views on science with the physics of today.

In this confrontation, we shall take two positions. First, looking across five centuries, we shall view the science of our era from the vantage point of Thomas, appraising those among the features of science that might have been of greatest concern to him; then, displacing ourselves to the present, we shall try to see the work of St. Thomas from amidst that larger pool of knowledge which has been gained in the intervening half millennium. While doing this, it is well to bear in mind that one of the subjects of confrontation, namely the philosophy of Thomas, is fixed aside from minor reinterpretation, while the other, contemporary science, is a moving, continually changing enterprise. Either perspective will therefore be altered as time goes on, and the conclusions reached today may not be valid tomorrow.

One thing, however, is clear beyond all doubt. The comparison of present scientific methodology with those aspects of Thomas' system that are applicable to science is more favorable today than at any time in the last two centuries, if not indeed more favorable than ever, in the sense of showing parallels and conjectures fulfilled in our epoch. Least favorable would have been a correlation at the end of the last century, when natural science was in a state of presumed finality and effective stagnancy, when physicists believed they knew the laws of nature in their essential form and were left only with the tedious task of determining the constants of nature with greater accuracy, when scientific philosophers of Häckel's camp proclaimed the doctrine of materialism as the last unalterable gospel.

All this has changed radically. Today, men of science are much less sure; the forms and essences of Thomas are not as strange to some of them as is Häckel's and Kelvin's ether. To be sure, scientists will not even claim for certain that their present more favorable judgment of the saint will stand the test of time. Yet most of us who believe in progress have at least a hope if not a conviction that the course of science converges upon or toward a goal of truth, and we find it hard to deny in our hearts that we are not a little closer to this goal than were our predecessors two generations ago. And if this hope or this conviction is justified, then the figure of Thomas has risen irreversibly to higher significance than it had before.

II

The central, perennial problem of the philosophy of science concerns the function and the reliability of reason and sense in human experience. When the accent is carried by reason the resulting philosophy is rationalism, idealism, phenomenology; when it rests on sense the result is empiricism, realism, positivism, sensationalism. The issue is far from settled today, for our halls of science still reverberate with the affirmation of physicists like Einstein and Eddington who embraced a distinctive a priorism; but their voices are commingled with protestations by biologists and psychologists who refuse to give speculation its due. In technical form, the problem is under current debate at the very forefront of researches in theoretical physics, where men are inquiring whether the proper stuff for theories of nature is to be the abstract form of mathematical reasoning or the texture of sensations that adhere to mechanical models of complex events.

The same problem forms the core of Thomas' epistemological studies. He states it very clearly and develops it in its historical setting in *Summa Theologica*, I, q. 84, a.6, from which I quote:[1]

Is intellectual knowledge acquired from sense-objects? No; that seems the answer, since intellectual knowledge transcends the things of sense. But to the contrary we have the teaching of Aristotle that the beginning of our knowledge lies in the senses. I begin my explanation by noting three differing opinions of philosophers on this question.

Democritus laid down that no other cause for any of our knowledge is required save the emission of bodily images from things and their entrance into our souls; the process of knowledge is an affair of images and discharges. The absence of any distinction between mind and sense underlies this opinion; the assumption is that all knowledge is like sensation, where a physiological change is induced by objects of sense.

At the other extreme Plato held that the intellect differed from the senses and was a spiritual power making no use of a bodily organ in its thinking. Now since a spiritual thing cannot be changed by a bodily thing, he was convinced that intellectual knowledge does not come about because the intellect is transmuted by sense-objects, but because it shares in intelligible and and bodiless ideas. Even the senses, he said, are powers acting apart from body; they are spiritual powers that are not transformed by sensible objects: though he allowed that the organs of sense undergo changes and immutations by which the soul is somehow stirred, and so formulates to itself the ideas of sensible things. Augustine touches on this opinion when he says that the body does not feel, but that the soul feels through the body, which it uses like a herald for announcing without what is expressed within. Plato's own conclusion is that neither intellectual knowledge nor even sensitive knowledge entirely derives from material things, but that sensible objects excite the sensitive soul to sensation and the senses likewise excite the intellectual soul to understanding.

Aristotle, however, took a middle path. He agreed with Plato that intellect differs from sense, but he also maintained that there is no proper activity of sense into which the body does not enter; sensation is not the activity of the soul alone, but of the body-soul compound. So also with regard to all the activities of the sensitive part. That sensible things outside the soul cause something in the human organism is as it should be; Aristotle here agrees with Democritus that the activities of the sensitive part are produced by the impressions of sensible objects on the senses — not, however, in the manner of a discharge, as Democritus had said, but in some other way. Democritus, incidentally, had also held that all action is the upshot of atomic changes. Aristotle, however, taught that the intrinsic activity of mind was independent of intercourse with body. No corporeal thing can impress itself on an incorporeal thing. Therefore the mere impression of sensible bodies is not sufficient to cause intellectual activity.

Thomas would, I think, have seen remarkable justification for his Aristotelianism in the methodology of present natural science. I speak of Aristotelianism following a custom which Thomas' modesty has permitted to be established. In truth, Thomism with its fine articulation, its incisive and systematic stand on matters that are vague in Aristotle, is very much closer and more relevant to modern science than the writings of any Greek philosopher. I shall try to make this apparent by sketching the epistemological situation in physics and its allied sciences today.

There are philosophers who still pretend that the problem alluded to and elaborated in the long quotation above is trivial, like most metaphysical questions. Among those detractors of metaphysics one finds first of all the Humeans who believe that in some manner all rational knowledge stems from sense, that rational principles lack certainty and universality, deriving their psychological force from prevalent but not necessary features of sensory experience. With this belief I shall not deal; much has been written concerning it, and there is hardly a modern physicist who is willing to accept it in view of the triumphs which highlight the recent history of science. To declare reason a secondary adjunct, a peripheral mentor or a record-keeping accountant of creative sense, such an attitude today encounters even emotional rejection; it marks an attitude no longer felt to be in need of careful philosophic refutation.

But the Humean has acquired a new friend in the logician who might put the case substantially as follows. It is clear, he admits, that reason is a powerful aid in the coordination and organization of sense data; science could certainly not do without it. Suppose, however, it had to renounce application of reason. Observation would then uncover facts and produce knowledge at a vastly slower rate; still, given sufficient time, the world would reveal itself as it does when reason is applied to the process of discovery. The categories and facilities of reason, though unique and not established, generated or conditioned by sensory knowledge, serve nevertheless merely to coordinate, correlate and unify the facts in a way that allows them to be dealt with logically under principles. While reason adds nothing of

its own to experience, it is the great sur-
veyor-from-on-high, the hawk's eye reg-
istering the movements of a host of empirical mice, or the complex electric com-
puter which handles and monitors regi-
ments of data. Whether these functions
are aptly described by Mach's ungenerous
phrase, according to which theory only
confers economy of thought, need not be
debated here; the point of the assertion
is that reason adds nothing of its own.

It is perfectly proper, then, to speak
of (a) an observational (or as I shall sometimes say, empirical or sensory) picture of
the world, and (b) a rational representation of that picture. Possession of all empirical details is tantamount to complete
knowledge of the world — provided the
mind can hold it. Rational knowledge, i.e.
a set of powerful propositions pregnant
with implications which coincide with the
sensory facts of experience, likewise span
the universe. We have, as it were, two
maps which correspond to each other: one

is an unwieldly and unperspicuous spread
of events in time and space, the other an
ever present, ever ready but involved set
of formulas. When a certain code is sup-
plied and logical techniques are adopted,
the latter can be translated into the former.

A human intellect, placed into the
world as a Lockean *tabula rasa,* could be
taught all there is to know in either one of
two ways. He could be told the facts and
be expected to draw his rational conclusions; or he could be acquainted with the
principles and laws together with initial
and boundary conditions so that he could
calculate what happened. If he prefers
the former treatment, he is an empiricist;
if the latter, a rationalist. The normal process of education selects a happy mean of
both procedures.

Now in view of this, what has become
of the metaphysical question of reason
versus sense? It seems to have lost its
force and its urgency if not its meaning.
For if one picture is equivalent to the

other, if the rational representation of knowledge is a mapping, an idle reduplication of what sense more clumsily conveys, what is the problem? Accent on reason pleases some minds, the straight facts are preferred by others. What appeared to be a profound problem is unmasked and found to be a preference between *languages*. Nor is there a conflict: reason can provide the whole language, and so can sense.

We were led to this conclusion by examining the existing body of experience, not experience in its genetic phase. How knowledge arises in the first body, whether every item of it first enters through the door of sense or whether there is also an ingress of novelty through reason — these questions are left unanswered by the preceding argument, although it should be noted that its advocates generally deny the second alternative. It is my own belief that novelty can enter through reason as well as through sense, but it seems unprofitable to move toward this more vital question before the representational problem concerning knowledge is more adequately resolved. Is it true that reason offers merely an image of the world of sense?

Fifty years ago, science's answer would have been affirmative without qualification. The reason is that theory was designed to permit prediction of observable phenomena in detail and that it succeeded in principle in doing this. And conversely, observations were fed as boundary or limiting conditions into the differential equations of physics, the result being an ever-growing, completely deterministic formalism capable of specifying or predicting every element of sensory experience. Failures were apparent to be sure, but these were ascribed to errors of observation or imperfections of the theoretical apparatus. There was nothing empirical which *in principle* could not be caught in the theoretical net, nothing theoretical

that did not have an empirical counterpart. The contrast between contingent existence and rational essence was dissolved,
one being an aspect of the other. There
was one reality in two representations.

The neatness of the scheme was occasionally drawn into question by workers
in the social and psychological sciences
who wondered whether human behaviour,
with its possibility of autogenous decision
and supposedly free will, could ever be
caught in the framework of a theoretical
formalism. In these sciences, success of
prediction had always come from statistical theories which renounced detailed description of phenomena. Thus in economics, *collective* effects like prices of goods,
stockmarket trends and so forth, became
of theoretical concern, not *individual trans-
actions.* Actuarial analysis was able to
predict the number of deaths in a group,
but not individual instances of death. The
psychologist felt he could anticipate a person's average reactions but not his re-

sponse to a specific situation. Such restriction of competence to statistical matters was explained in social science as in meteorology, by the complexity of the subject matter, but it was believed that ultimately, when a perfect science is at hand, specificity will be part of perfection. For was not physics, the most highly developed of the natural sciences, even then able to formulate its laws in a manner adequate to detail? Did it not forecast every feature of the motion of a single stone, the position of a star, or the outcome of a specific reaction?

Quantum mechanics has changed all this. Collective analysis, the use of probabilities which was thought to be enforced by the complexity of the social phenomena was required in the description of the most elementary, the simplest facts as well. The motion of the smallest particles was found to be subject to the laws of large numbers, any individual instance showing evidence of intractible caprice. No finite set of observations enables theory to predict the place where an electron will strike a screen; quantum mechanics restricts itself to the calculation of averages, of possibilities indicating how many times in a great number of passages the electron will hit a given spot. More than that, it gives convincing reasons why this resignation is necessary and why the detailed determinism of classical physics must fail.

Let me forego the belaboring of these new results of physical science — the uncertainty principle, the failure of dynamic causality and the rest. They are too well known to philosophic audiences to merit exposition. What needs stressing is their effect upon the age-old problem of reason's relation to sense. For the new developments mean that the sensory domain is forever richer than theory since individual sensory events have no direct counterpart in theory, and on the other hand, it is clear that the concepts of reason (such as state functions and probabilities) cannot be directly sensed. In the realm of the atom, the isomorphism between the rationalistic image and the sensory image of reality is broken, one is not a mere duplication of the other, the relation between them is not an idle sort of identity; they are not equivalent languages since things can be said in one that cannot be said in the other.

If I may be permitted to use a terminology employed in another publication, what appeared in classical science as a homogeneous picture describable arbitrarily either in terms of particular data or in terms of universal laws, has undergone a basic fission into two parts: historical and physical reality. The former comprises individual ("P-plane") data, details of introspective awareness, personal decisions and all the vivid existential features that make up history; the latter is the sum total of all lawful experiences, the principles and laws of science, the trajectories of physical bodies as well as the space-time behaviour of the state functions which represent atomic systems. To be sure, in our ordinary, common sense experiences these two realities, these two partial aspects of our cognitive experience, still largely overlap, but the fundamental break occurs in the domain of the very small, where physics has made its most recent advances.

The relation between historical and physical reality is no longer that of a simple isomorphism. This means that the relation between sense and reason is no longer trivial or, in the saint's phrase, "The mere impression of sensible objects is not sufficient" for the intellect.

Thomas goes on in a remarkable passage which is seen to be truer today than at any earlier time after it was written. He says:

A nobler and higher force is required, for the agent is more honourable than the patient. Not, however, in such a way that intellectual activity is caused in us by the sole influence of some higher being, as Plato held, but that there is a spiritual ability within us, called after Aristotle the *factive intellect,* which by the process of abstraction renders actually intelligible images taken from sense. As regards these images intellectual knowledge is caused by the senses. Yet because an image alone is not sufficient to transform the receptive mind until it has been heightened and made actually intelligible by an *active intellect,* we should not say that sense-knowledge is the total and perfect cause of intellectual knowledge. Let us say instead that sense-data are by way of offering the material for the cause.[2]

Since sensuous and intellectual being are poles apart, the material form is not immediately accepted, but is first worked up and spiritualized by an elaborate process and so brought to the reason.[3]

Factive intellect, to use a phrase coined by Thomas Gilby in translation of Aquinas' *"intellectus agens"* and Aristotle's γοῦσ ποιητικὸς, this happy wording calls attention to an element of epistemology, an autonomous function of reason, which has long been buried under layers of positivistic and language-analytic considerations. In the philosophies that accompanied the science of the last century, and in many doctrines prevalent today, reason is accorded the role of comparing, sifting, correlating and organizing empirical facts; its highest function is abstraction, distillation from experience of certain formal features present within that experience but somehow hidden from view. Science is said to be inductive, and it is the nature of induction to prohibit reason from injecting into the stream of knowledge any elements of its own. By a lapse of caution such an injection does occasionally occur, but it is the scientist's duty to search them out, to recognize them as figments (sometimes called "mere" constructs, or constructs of the imagination) and to eliminate them from scientific discourse.

A study of procedures employed in the macrocosm seems to bear this out. As long as interest is confined to matters directly accessible to ordinary sensation, observation and geometrical analysis, concepts are abstractions from sensory data. The concept "tree" has in a fairly convincing sense, no ingredients which do not point to aspects of sensation, and it may be said that the three-dimensional object outside my mind, outside my window, represents, aside from the visual features of present awareness, nothing but abstractions from previous or at any rate possible sensations. A difficulty arises, it is true, with respect to the imputation of permanence to the tree, its property of being out there when I am not looking, for it is hard to see how an unsensed presence can possibly be related to sensation. However, there are ways (all of them impassable to me) which take one round this one isolated difficulty.

The situation is quite different when a study of the atomic microcosm or, for that matter, anything removed from direct sensation, is coupled with the ordinary appearance of things. Eddington called attention to this point when he discussed the "two desks" at which he sat and worked. One of them was the macroscopic desk with its geometric shape, its smooth surface, brown color, its sensed rigidity and its inert mass. The other was the "desk of the physicist," consisting of interlocking space lattices of different molecules, each molecule containing its atoms and each atom its nucleons and its electrons circulating with tremendous speeds. But this is not the whole unpleasant story. According to the laws of quantum mechanics, the electrons may not be pictured as rapidly moving small o b j e c t s having definite speeds and orbits, for they do not have definite positions at all times and their behaviour is not properly accounted for by the pictorable trajectories of ordinary things in space. Rather, to use a qualitative simile from Thomas, they display the caprice of angels of whom he says: "Motus angeli potest esse continuus et discontinuus

sicut vult. . . . Et sic angelus in uno instante
potest esse in uno loco, et in alio instante
in alio loco, nullo tempore intermedio
existente."[4]

While this is an amusing rendering of
the uncertainty principle it is of course not
wholly correct even as an analogy. For
where St. Thomas leaves us free to visual-
ize, to intuit in terms of a dicontinuous
model of motion, the passage of an angel,
the electron should not be visualized at all;
it should be described in abstract mathematical terms which suppress, in general,
the reference to a specifiable position. And
this is by no means absurd, for the electron
in contradistinction to the desk is so small
and elusive that no direct sensation could
ever apprehend it. Let it also be noted
that this is not a contingent infelicity aris-
ing from the crudeness of our sensory
equipment. If the present form of quantum
theory is correct, then a contradiction
arises within the theory itself when the assumption is made that the electron can be

perceived in the manner of a moving stone. This, then, confronts us with Eddington's dilemma. The second desk, the physicist's (or better, the mathematician's) desk, has ultimate elements which cannot be perceived. He who identifies the two desks is guilty of saying that an object wholly given in sensation can be decomposed by the magic of science into constituents not given in sensation. Reference here is not to the trivial difficulty involved in the assertion that a large visible object may be subdivided into parts too small to be seen; this we understand in terms of the resolving power of our eye. The troublesome feature in our present example is that desk No. 1 has property A while desk No. 2 has property B. To say that the desks are logically identical requires an understanding of A in terms of B and such understanding is wholly lacking.

Our analysts have tried to persuade us with great verbal skill to believe that there is no dilemma and have accused Eddington as well as pedestrians like myself of linguistic carelessness. Still I fail to see how any amount of verbal therapeutics can cover up the patent fact that Eddington's two desks are *not* the same and never will be. They differ in this: the second desk is the product, very largely, of Aquinas' factive intellect; the first involves mere abstraction, collation of sensory material, performed by what Thomas calls the passive intellect. The two desks are not identical; they are related by a relatively new epistemological factor which I have termed a rule of correspondence in p r e v i o u s writings. It is a link between sense and reason, a relation often mistaken for an identity but exposed as different from identity by recent science and recognized as important by Kant ("schemata"), Reichenbach ("Zuordnungsregeln"), C a r n a p and Northrop ("epistemic correlations").

The difference between Eddington's two desks appears in every careful analysis of all scientific quantities or objects. It distinguishes the force conceived as a push or a pull from Newton's "mass times acceleration;" the radiation felt by the hand from the electromagnetic wave passing through space; the thermometer reading from the mean kinetic energy of molecules; a sensed interval of time from the sort of quantity that contracts in another inertial system; the taste of an acid from its pH concentration; the felt pain from a physiological state of the brain; the count of three from the number three. In each of these examples, one entity is a construct generated by factive reason, its counterpart a set of percepts devoid of sufficient rational cohesion to make scientific treatment possible. There is a sense in which the constructs are "suggested" by the percepts, but very little is gained by the use of that indefinite phrase which merely adverts to the circumstance that when the perceptory situation is present we are urged by psychological predisposition or methodological habit to accept the presence of the other. The difference, the relation in question, is not one of the well known and extensively studied relations of logic; its nature can be exhibited only by a study of the process of knowledge which shows it to be a unique relation between ostensibly different kinds of experience. And the differentiating feature in experience is the function of Aquinas' factive intellect in one but not the other.

The preceding statement, is, I fear, a bit too strong, for it might lead to an objection put to me very kindly by Fr. Gerard Smith, who anticipated what will doubtless be the attitude of my critics. The factive intellect's function is precisely the same, he affirms, both in judgments of reason and in the judgments of sense perception, viz., the role of making actually intelligible *any* potentially intelligible object, whether that object be sensed or understood.

The difficulty thus raised disappears perhaps when attention is drawn to two things which have been assumed but left unsaid. First, it must not be supposed that there exists a sharp distinction between factive and passive reason. The elements of our experience cannot be classified in static fashion; they move elusively across all boundary lines which the logician or the psychologist may draw for them. No actual determinate experience is, for instance, wholly perceptory or wholly rational; it may be strongly accented on the perceptory or on the rational side. Nor is a given act purely cognitive or purely emotional; it is at best a very unequal mixture of these two categories of comprehension. And so it is with the faculties that have been assigned to the mind for dealing with the ever-blending flux of experience; sense and reason are polar extremes of activity within the cognitive process; they span a vast domain of functions that are neither one nor the other but partake in differing measures of the characters of both. Clearly the same is true of the distinction we have made within the category of reason, viz., between the factive and the passive intellect.

The second point concerns the matter of judgment in sense perception. Pure perception, it seems to me, is an idealized, a limiting kind of experience which one must deliberately abstract from a perceptive act. It is a wholly uninterpreted awareness, a complex of sensations not yet reified. Robert Penn Warren's novel, *All The King's Men*, which is extremely rich in its description of the existential features of unclassifiable experiences, contains an illustrative passage which eloquently verifies this claim.

> "Lucy Stark looked up at the Boss right quick, then looked away." Her husband tries to interpret the look but fails. Warren goes on to say: "It didn't have any question, or protest, or rebuke, or command, or self-pity, or whine, or oh-so-you-don't-love-me-any-more in it. It just didn't have anything in it, and that is what made it so remarkable. It was a feat. Any act of pure

perception is a feat, and if you don't believe it, try it sometime."

In the strictest sense, therefore, there is
no judgment concerning pure perceptions.
When perceptory features are discrimi-
nated, collated, related as in a judgment
pointing to a thing which the perception
implies, then clearly, factive reason is already at work; but we have transcended
the domain of perception in its typical
sense.

But how is the construct, the product
of factive reason, related to the thing it
apprehends? Here again, the nucleus of
the answer may be found in Thomas, albeit
expressed in the scholastic language of his
day and in terms of the science accessible
to him. As I read it, I find it anticipating
in essence a theory with reference to
physics which I published 23 years ago.[5]
Its gist is the following:

The constructs, invented by reason as
counterparts to certain impressions of sense
(P-plane of experience) and related to

them by rules of correspondence much in the manner in which a mathematical group is mapped upon one of its representations, are subjected to two kinds of validating procedures. The first of these procedures involves abstract principles (called metaphysical requirements) which do not have their origin in the sensory part of our experience but spring from what Thomas would call the nature of man's rational soul. The second set of validating procedures involves a mutual reference between reason and sense (C-field and P-plane) via the assumed rules of correspondence and is called, loosely, empirical confirmation. A further elaboration of these matters is hardly in place here, for it would introduce elements into our analysis which are of recent date, relating to very modern science, and could not possibly have been foreseen in their detailed specificity by Thomas. There is, however, one feature that seems worth recording.

The metaphysical rules, when seen in retrospect, appear to change as man's understanding of the universe and of himself progresses. I have found no explicit indication of this slow alteration, this gradual unfolding of the nature of man's intellection in the writings of the saint, nor did I discover arguments against it. It is clear that a rigid and final fixation of the metaphysical principles underlying science is neither necessary for the possibility nor exhibited in the course of science: static and universal categories of the Kantian type are not easily reconciled with the present mood of research. Thomas' view appears to be more tolerant; at any rate, his denial of conflict between natural science and revelation may be taken to sustain the gradual approach to finality, through incidental stages encumbered by error, of metaphysical reason.

Constructs satisfying the principles of reason and also confirmed by empirical tests may be called verifacts. And the verifacts, or at least a certain definable class of them, compose within our experience the domain of physical reality.

This is the end of the *epistemological* story, in which science is the primary actor. It leaves reality within experience and therefore stops short of Thomas' ontological position. I note this with emphasis, for it seems to me that beyond this point science is no longer a reliable guide, except for him who believes the naive claim of textbook writers that the first principle of science is to recognize an external world. This, of course, is an unprincipled assertion if not perfect nonsense, for even the slightest inquiry into the meaning of this claim exposes its faults: science does not reveal a world outside experience in terms of any knowable properties that can be said to adhere to it, nor does it, for its own purposes of understanding, require it.[6]

At the same time most scientists do believe that experience points to an ontological reality beyond the physical which

consists only of verifacts. To reach ontological reality he must make a leap, a commitment of a kind transcending, I believe,
those which enabled him to be a scientist.
He must indeed entrust himself to other
hands.[7] And here, it seems to me, Aquinas'
doctrine of essences, his theory of the scale
of being of substances serve useful purposes. While I am unable to accept all of
his details, the systematic structure of the
doctrine is impressive.

Doubtless you are tempted to ask: if
science finally makes the transition from
reality within experience to ontological
reality, why not perform the ontological
leap at the beginning? For the simple reason that one should not jump into a lake
before one has learned to swim. Scientifically uninformed ontological commit-
ments have all too frequently impeded the
progress of science, destroyed the apti-
tude of the person who makes them to pur-
sue scientific tasks. They have prejudged
the results of science, conferred on com-

mon sense an undue measure of plausibil-
ity and given popular attitudes an excess
of inertia which science struggles valiantly
and sometimes vainly to overcome. One
senses a little of this caution in Thomas
when he argues against Parmenides,[8] saying "His mistake lay in treating the nature
and meaning of being after the fashion of
a generic object. But being is not a genus,
it is predicated in manifold ways of different things."

Let us return to the epistemological aspects of reality and consider once more
the constructive role played by factive reason. In *Summa Theologica* I, q.85, a.2 we
read:[9]

> Some have held that our cognitive
> powers know only the impressions made
> on them, for instance, that sense knows
> only the alteration of its organ. According
> to this reading, mental states are the ob-
> jects of knowledge.

What Thomas here says is that the verifacts, originally introduced by the con-

structive power of reason, rise above the
status of mere mental images and take on
an objectivity of their own. There is an
aspect in which they are mental states, for
they can always be recognized as such
when reason reflects upon itself; yet their
character does not exhaust itself in this
peculiarity since verifacts refer objectively
to other constructs, are capable of relations
that are not experienced as mental, including their relation to mental states.

Potent arguments are given against the
view which identifies constructs with mental images:

> This opinion seems to be false for two
> reasons. First, if the objects of intelligence
> and science are merely psychic conditions,
> it would follow that science does not deal
> with non-mental things, but merely with
> impressions in consciousness.
>
> Secondly, it would revive the ancient
> error of maintaining that whatever seems
> so is truly so, and that contradictories are
> simultaneously tenable. Were a cognitive
> power able to perceive no more than its

own proper states, then it could judge only about those. Now an object appears according to the manner the faculty is affected, and were modes of consciousness the only data, a faculty could judge merely its own proper impressions, and every judgment would be true; when a clean tongue judges honey to be sweet then it would judge truly, and when a dirty tongue judges it to be bitter then also would it judge truly, in each case "going on its impressions. Every opinion would be equally valid; so also in general would be whatever was fancied."[10]

Before closing this section on modern science, as viewed from the station in history occupied by Thomas, I cannot forego a few more qualitative, brief considerations. Physics has grown increasingly abstract and its reliance on formal, often mathematical principles, has been strengthened. The force of relativity springs from its postulate of invariance, quantum mechanics features principles of symmetry, and all these result in instances of empirical veridicality that are amazing. No one who reads St. Thomas and feels his devotion to neatness of argument, simplicity and elegance of exposition, can doubt that he would be pleased at such developments in recent physics.

But there is one element in it which might make him frown. The discovery of non-Euclidean geometrics and their use in physical explanation, the discovery of many different systems of logic including those which affirm a *tertium datur,* together with many of the abstract pursuits just mentioned, have deprived us of confidence in a priori proofs. The conclusion seems inescapable that no proposition, formal or otherwise, carries within itself complete assurance of its truth. It appears to me that this stands in contrast with *XI de Veritate 1,* where we read:

> If someone proposes statements that are not developed from[11] self-evident principles, or if he fails to make the connection clear, he does not cause scientific knowledge, but opinion, or perhaps faith.[12]

In the passages that follow it seems that Thomas takes his innate principles to be chiefly the laws of logic. If restriction to these were intended, a large number of contemporary scientists would follow him.

Interesting things have happened with respect to the doctrine of creation. While the *Summa Theologica* is very refined and non-commital in developing the theory of creation, and while the word is taken in a wider than liberal sense, there is none-theless the statement in *De Potentia*: "that God can and does make something from nothing should be steadfastly held."[13]

Such an affirmation was anathema to the scientists of the last century, at least so far as matter is concerned. Relativity has changed all this, and it is a curious fact, perhaps not widely known, that cre-ation of matter out of nothing contradicts no physical conservation law. In oversim-plified form the argument runs as follows:

If the universe were empty its energy would be zero. Let us call this state of nothingness S_o. For comparison we consider a universe filled with a spherical aggregation of matter to be designated the state of existence, S_e. If the total mass in S_e is M and its radius R, the gravitational potential energy of S_e is the negative quantity $-k\,G\,\dfrac{M^2}{R}$, provided G is the constant appearing in the law of universal gravitation and k a numerical factor not greatly different from 1 and depending upon the exact distribution of matter within the sphere. But because of the mass-energy equivalence, the aggregation of matter also contains the positive energy Mc^2, c being the speed of light. Hence the total energy of S_e is: $\left| \; E_e = Mc^2 - kG\,\dfrac{M^2}{R} \right.$ whereas $E_o = O$.

Creation is energetically possible if $E_o = E_o$, that is if $E_e = O$, and the latter condition is fulfilled when $\dfrac{kGM}{Rc^2} = 1$

The remarkable fact is that this relation may well be fulfilled by our actual universe. To be sure, knowledge of the constants involved is far from precise, and the uncertainties attach primarily to M, the total mass of all stars, and R, the radius of curvature of the (finite) universe. But the following values seem to be among the best available:

$G \,/\, c^2 = 10^{-28}$ cm gm^{-1}, M $= 10^{55}$ gm, R $= 10^{28}$ cm. With these data k needs to be about 10 in order that $E_e = E_0 = 0$. One truly wonders whether this is an accident!

There is no difficulty in satisfying the other conservation laws connected with this process of "creation," since linear and angular momentum are 0 for both S_0 and S_e. Let it be acknowledged, however, that the simple computation here given ignores many facts: it omits the mass of dust in interstellar space, the energy of radiation and of cosmic rays, all interactions except the gravitational attraction. Yet it is unlikely that these effects will alter the result significantly.

III

In the foregoing section we tried to listen to the voice of Thomas, to several of his specific pronouncements, and examined them for their prophetic quality in regard to recent science. Here we shift the emphasis; we proceed to outline first the philosophically decisive features of science and then reflect, in less enumerative or itemizing fashion, upon the stature of Aquinas as a philosopher of science.

The approach in this section prompts one to offer an apology, for I shall at first describe the method of science without r e f e r e n c e to our philosopher's own writings. If the initial comments seem too discursive and unrelated to the proper subject under discussion, it is my hope that the similarity between the view I sketch and the basic attitude of Thomas will emerge more effectively in the end.

Many of us, especially the scholars in the liberal arts and nearly all newspapermen, still look upon science as a catalogue of dead but certified facts. Our teaching largely reflects this view. A first course in physics, chemistry or biology has come to be identified with a certain subject matter, a collection of facts sometimes unilluminated by ideas, hanging together as a traditional aggregate and forming a whole which is exactly the sum of its parts. When told that a student has had half a course in elementary physics one knows accurately what this means. And one also knows that the course contained no ideas, for ideas, in contradistinction to facts do not obey the laws of arithmetic. Laws are presented as inductive generalizations of observational experience, never as bold conjectures, as ideas defying the knowledge of their day, which is, indeed, the status wherein most of them were born. One reason for the shortage of scientists in America, I believe, lies in the extinction of interest resulting from that massive embrace with facts which is the essence of almost every beginning science course.

Scientific research, too, is widely supposed to be the uncovering of new factual knowledge, preferably without interpretation, i.e., without commitment on the part of the scientist as to its meaning. It turns up rocks to see what is under them, and it then proceeds to describe carefully, honestly and without shame. When somebody discovers how a primitive tribe, forgotten except for the study in question and entirely without significance in history, tilled its fields and educated its young, we are expected to be tremendously impressed; statistics concerning the sexual habits of unspecified human males and females are supposed to leave us scientifically speechless. All this because discovery, not quest for meaning, is the heart of science. Public recognition of greatness in science has all too often singled out the discovery of new elementary particles, chemical elements, useful compounds, ignoring the creation of powerful new ideas.

Nor is the factual emphasis restricted to the natural sciences. Historians, wishing to attain complete objectivity, sometimes speak of factual reporting as the true style of historical writing and condemn interpretation as biasing or falsifying the account. The authority of Ranke can be cited in favor of this attitude. Fortunately, there are now men like Spengler and Toynbee who attempt to see meaning and purpose in the course of history, and to me, these resemble the modern scientist more closely than any of the historical reporters of the last century. In science, facts are fertilized by ideas; I wish our culture took the historical analogue of this proposition more seriously: communism shows how powerfully persuasive, how sinister if erroneous, a coupling of an ideology with historical reporting can be.

To the view of science here criticized, research is like the solving of a picture puzzle. Piece after irregular piece, fact after peculiar fact, is found and fitted into place, and when the pattern is completed the scientist's job is done. The whole world is such a picture puzzle, and there are optimists who believe that some day the universal puzzle will be solved and the happy millennium will begin. The trouble with this simile is easy to recognize. It fails to account for the progressive character of science in mistaking it as accretion of facts; it ignores the most crucial point that a scientific problem is never completely solved, for in terms of the puzzle, research is not confined to the field outside the completed pattern but continually reforms and alters the picture which was thought at an earlier time to be already finished.

The search of the picture puzzle fan is for information, for time-bound knowledge. But science conveys understanding as well. And understanding differs from knowledge in a manner suggested by the word itself. To understand implies a third dimension below the surface of the facts. Here we find the texture of concepts, laws and principles which confer significance upon the changing surface pattern of the facts, we find the fertile soil in which the facts are rooted. Here is the ground of meaning, scientific and otherwise, for meaning always rises from below.

The wise teacher of science knows this well and uses the illuminating, integrating and dynamic effects of ideas in his classroom. Rather than collect laboriously, the painful empirical evidence which led, for instance, to the law of energy conservation he will, in a spirit of adventure, suggest Joule's and Mayer's exciting conjecture as a captivating idea and then verify it instance by instance. The facts will then be more than facts; they will be occasions with significance, memorable and welcome.

Were I to propose an allegory for the scientific process in place of the picture puzzle I would choose a growing crystal. Imagine a substance at a temperature just above the melting point. Every element of liquid moves in random fashion, and if the pattern of the motion is made visible it is observed to have the existential beauty of irregularity and caprice. Very little can be predicted concerning it in detail. The motion of one small portion of liquid is almost uncorrelated with the remainder; there exists only what the physicist calls short-range order, a correlation function falling to zero beyond the second or third neighbor to a given molecule.

Now let the temperature fall below its melting point, so that crystallization begins. Somewhere, usually at an unexpected place within the liquid, a little nucleus will form and grow into a pattern of geometric symmetry, regularity and beauty. Within this pattern everything is clear, orderly and predictable; yet the substance is unmodified, its chemical character is unchanged even though its appearance to the eye is altered. If the original liquid has a sufficiently large volume and its boundaries are far away, the growth of the crystal takes a peculiar course: here it advances along a broad front, there it stretches a narrow arm into the molten mass, and elsewhere it progresses in unsteady, trying fashion. Where the next advance occurs is difficult to foresee in general. There is, however, a way to stimulate the growth by artificial intervention, and this occurs when certain impurities or, still better, small crystalline fragments known as seeds, are inserted into the liquid.

I think of human experience in its primary form as the amorphous liquid mass, of science as a crystalline growth which imparts order to the shapeless mass. Experience must not, of course, be taken in the narrow positive sense of external or

sensory experience impressed upon the
word by the British empiricists, but in its
elementary richness still preserved in an
unreflective usage of the word, which the
dictionary defines as "the sum total of con-
scious events which compose an individual life." Not only sensations, but also
feelings, thoughts, decisions, actions and
even dreams compose experience. The
range of experience, the volume of the
amorphous mass, is in principle unlimited,
for no one can say what novel features
may arise within our lives from the fertile
ground of being under circumstances yet
unforeseen. As does the liquid, experience
in this primary prescientific phase contains only short-range order, contains but
meager facilities to predict. It is not, however, for that reason less important or
ignorable; in fact it is the substance of
which the larger part of our lives is made;
it fills the domains vaguely formalized as
ethics, politics, history and religion, com-
poses a large part of sociology, economics

anthropology and psychology; it is present in the so-called biological sciences and, in a certain measure, in the physical sciences as well. The most momentous decisions must, perhaps unfortunately, be made within this amorphous experience where short-range order prevents us from anticipating their long-range effects. The philosophy adequate to the character of primary experience is existentialism, and were it not for the regularizing consequences of man's scientific propensity, or shall I say destiny, which is likewise part of experience, existentialism would stand aright.[14]

Within amorphous experience, t h e crystal of science grows. Its natural growth is unpredictable, although means exist for speeding and directing it by insertion of crystalline "seeds." The ordered structure now extends through much of physics, chemistry, some parts of biology, but even in these fields it has great holes. Growing edges are visible in economics, psychology and social science, indeed in history, ethics and religion. The growth is not malignant, for it never adulterates the substance it informs. A light of understanding, a satisfying order enabling prediction pervade the organized lattice of scientific experience; none of the essential qualities of primary experience is taken away but depth and illumination are added.

There is no end to the process of growth. The shapeless matrix of our experience is unlimited, and the crystal will never span it all. That millennium of the picture puzzle devotee, the golden age when the puzzle can be said to have been solved, will surely never come, and he who reserves his reverence for the uncrystallized part of his experience is in no danger of losing it. What I foresee is an infinite crystal growth into an amorphous domain whose volume is likewise infinite yet vastly greater. Because of this the existential-

ist will probably never cease to quarrel
with the rationalist.

But all parables are imperfect, and the
one here sketched suffers from one major
fault. It pictures experience as a large
volume filled with substance first un-
formed, then organized, but as if the total
mass of substance were constant. In our
experience this is far from true. New elements of substance are created every-
where, its mass and volume constantly
increase. And so the crystal must be
thought of as continually developing holes
and crevasses filled with liquid, which are
then incorporated into the orderly structure. Occasionally, however, such incor-
poration does not succeed and a large part
of the crystal melts again, later to re-form
itself after a new and more inclusive pattern.

Thomas' figure, when viewed through
the vista of these considerations, appears
almost in an attitude of benediction upon
their truth. For he above most others en-

deavored to put systematic structure into the shapeless mass of human experience available in his day. The fact that he included religious experience or, if you will, placed *major* emphasis upon religious experience, cannot alter this appraisal; it means that the trajectory of his teaching carries beyond our time. I like this dynamic quality in Aquinas' message, and I regret the petrefaction achieved by the monument builders who admire nothing but the fixed structure of his system of thought, who invest him with an aura of final and eternal truth. The saint himself ascribed such truth to God, holding that man must search for it in indirect derivative fashion, for he said, in *Summa Theologica* (I, XIV, ad 3):[15] "Natural things lie midway between God's knowledge and ours. Human science derives from them, and they derive from God's own vision."

NOTES

1. All quotations from St. Thomas included in this lecture are in *St. Thomas Aquinas: Philosophical Texts*, selected and translated by Thomas Gilby, London, New York, Toronto, Oxford University Press, 1951. Cf. #650, pp. 241-243.

2. *Sum. Theol.* I, q.84, a.6; Gilby #650, p. 243.

3. *Quaest. Disp. de Anima*, 20; Gilby #651, p. 244.

4. *Sum. Theol.* I, q.53, a.2.

5. H. Margenau, "Methodology of Modern Physics," *Philosophy of Science Journal*, II (April, 1935) 48 and 164. Cf. also *The Nature of Physical Reality*, New York, McGraw-Hill, 1950.

6. Lest the reader be puzzled, it should be said that the word "experience" in this context (and in all my writings) does not designate that curtailed and truncated sensory activity which most empiricists mean by it. Nor is the verb "to experience" necessarily transitive.

7. My chief reason for saying this is that within the realm of epistemology empirical verifaction is possible; in the territory where the "leap" lands us, it is not.

8. *Com. in I Metaphys.*, lect. 9; Gilby #429, p. 149.

9. Gilby #595, p. 218.

10. *Sum. Theol.* I, q.85, a.2; Gilby #595, p. 218.

11. *Includuntur:* the point here made is unaffected even if a more literal translation is preferred to Gilby's.

12. Gilby #1091, pp. 379-380.

13. *Quaest. Disp. de Potentia* III,1,6; Gilby #390, p. 132.

14. Father Smith suggested that I amplify these rather cryptic allusions to existentialism. This is difficult to do in a brief space, particularly because of the many different forms which that doctrine has taken. Leaving aside all aspects without direct bearing on epistemology, and ignoring especially the sordid, the pietistic and the heroic claims made by the advocates of the view, the following may be said as a characterization of my own position.

The existential philosopher's prime concern is with existences, bare facts, unpredictable contingencies, which he tends to oppose to the essences, the regularities and the laws of science, the latter being regarded as futile in human affairs. Now the status of any such view *vis à vis* science underwent a distinct change as a result of the development of the quantum theories in modern physics.

In pre-quantum days, existentialism was merely an acceptance, often too ready an acceptance, of the limitations of man's intellect. The competence of science was correctly portrayed in the famous dictum of Laplace, who said: "An intelligence knowing, at a given time, all forces acting in nature as well as the momentary positions of all things, would be able to comprehend the motions of the largest bodies and those of the smallest atoms in one single formula, provided it were sufficiently powerful to subject all data to analysis." In principle, and strictly speaking, then, there are no brute unpredictable facts: contingencies arise from ignorance and from imperfections of our intelligence. Necessary imperfections and inevitable ignorance, to be sure, are reason enough to justify a large part of existentialism's claim, and this is why I suggested that the doctrine will always have relevance in the amorphous sphere of human experience.

But quantum mechanics has given more substance to the philosophy under review. For it has falsified Laplace's proposition by giving impressive evidence that even a perfect intelligence (Laplace's demon) is unable to comprehend or predict all individual events. Existentialism need no longer base its claim on human imperfections; it can point to the very foundation of scientific theory and find the substance for irreducible existences at the fountain head of underdstanding. To my knowledge, this fact has never been used by existential philosophers to their advantage. In so far as existentialism makes closer contacts with ethics and religion than does science proper, there is perhaps some hope that existentialism might form a link between science and religion.

In all honesty, however, one must add that much current existentialism vastly overplays the good hand it holds; for it forgets that there remains a vast domain of experience in which essence binds existence, i.e., in which facts have perfectly regular and non-contingent relations to a known structure of law and order; and it minimizes the role of man's intellect at a time when the only prospect of human survival should marshall all intellectual resources. Extreme existentialism at the present moment is not so much erroneous as it is tragically unfortunate.

15.   Gilby #45, p. 18.

The Aquinas Lectures

Published by the Marquette University Press,
Milwaukee 3, Wisconsin

St. Thomas and the Life of Learning (1937) by
the late Fr. John F. McCormick, S.J., professor of philosophy, Loyola University.

St. Thomas and the Gentiles (1938) by Mortimer J. Adler, Ph.D., associate professor of the
philosophy of law, University of Chicago.

St. Thomas and the Greeks (1939) by Anton C.
Pegis, Ph.D., former president of the Pontifical Institute of Mediaeval Studies, Toronto.

The Nature and Functions of Authority (1940)
by Yves Simon, Ph.D., professor of philosophy of social thought, University of Chicago.

St. Thomas and Analogy (1941) by Fr. Gerald
B. Phelan, Ph.D., professor of philosophy, St.
Michael's College, Toronto.

St. Thomas and the Problem of Evil (1942) by
Jacques Maritain, Ph.D., professor *emeritus*
of philosophy, Princeton University.

Humanism and Theology (1943) by Werner
Jaeger, Ph.D., Litt.D., University professor,
Harvard University.

The Nature and Origins of Scientism (1944) by John Wellmuth.

Cicero in the Courtroom of St. Thomas Aquinas (1945) by the late E. K. Rand, Ph.D., Litt.D., LL.D., Pope professor of Latin, *emeritus,* Harvard University.

St. Thomas and Epistemology (1946) by Fr. Louis-Marie Regis, O.P., Th.L., Ph.D., director of the Albert the Great Institute of Mediaeval Studies, University of Montreal.

St. Thomas and the Greek Moralists (1947, Spring) by Vernon J. Bourke, Ph.D., professor of philosophy, St. Louis University, St. Louis, Missouri.

History of Philosophy and Philosophical Education (1947, Fall) by Étienne Gilson of the *Académie française,* director of studies and professor of the history of mediaeval philosophy, Pontifical Institute of Mediaeval Studies, Toronto.

The Natural Desire for God (1948) by Fr. William R. O'Connor, S.T.L., Ph.D., former professor of dogmatic theology, St. Joseph's Seminary, Dunwoodie, N.Y.

St. Thomas and the World State (1949) by Robert M. Hutchins, former Chancellor of the University of Chicago.

Method in Metaphysics (1950) by Fr. Robert J. Henle, S.J., dean of the graduate school, St. Louis University, St. Louis, Missouri.

Wisdom and Love in St. Thomas Aquinas (1951) by Étienne Gilson of the *Académie française,* director of studies and professor of the history of mediaeval philosophy, Pontifical Institute of Mediaeval Studies, Toronto.

The Good in Existential Metaphysics (1952) by Elizabeth G. Salmon, associate professor of philosophy in the graduate school, Fordham University.

St. Thomas and the Object of Geometry (1953) by Vincent Edward Smith, Ph.D., professor of philosophy, University of Notre Dame.

Realism and Nominalism Revisited (1954) by Henry Veatch, Ph.D., professor of philosophy, Indiana University.

Imprudence in St. Thomas Aquinas (1955) by Charles J. O'Neil, Ph.D., professor of philosophy, Marquette University.

The Truth That Frees (1956) by Fr. Gerard Smith, S.J., Ph.D., professor and director of the department of philosophy, Marquette University.

St. Thomas and the Future of Metaphysics (1957) by Fr. Joseph Owens, C.Ss.R., associate professor of philosophy, Pontifical Institute of Mediaeval Studies, Toronto.

Thomas and the Physics of 1958: A Confrontation (1958) by Henry Margenau, Ph.D., Eugene Higgins professor of physics and natural philosophy, Yale University.

Uniform format, cover and binding.

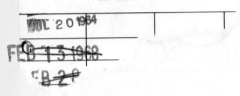